PLANETARY EXPLORATION
THE MOON
Don Davis and David Hughes

Series editor: Dr David Hughes

PLYMOUTH SCHOOL
919 Tollcross Road
North Vancouver, B.C.
V7H 2G3

Facts On File
New York • Oxford

The Moon
(Planetary Exploration V.2)

Copyright © 1989 by **BLA Publishing Limited**
All Don Davis artwork © **Don Davis**
All rights reserved. No part of this book may be reproduced
or utilized in any form or by any means, electronic or mechanical,
including photocopying, recording, or by any information
storage and retrieval systems, without permission in
writing from the publisher. For information contact:

Facts On File, Inc.
460 Park Avenue South
New York NY 10016
USA

Library of Congress Cataloging-in-Publication Data

Hughes, David W., Dr.
    The Moon / text, David Hughes ; illustrations, Donald Davis.
      p.   cm. -- (Planetary exploration)
    Includes index.
    Summary: Discusses the latest data available about our Moon,
including a review of conditions there which could support life.
    ISBN 0-8160-2046-9 (alk. paper)
    1. Moon--Exploration--Juvenile literature.   2. Moon probes-
-Juvenile literature.   [1. Moon.]   I. Davis, Donald, ill.
II. Title.   III. Series.
QB582.H84   1989
523.3--dc20   89-31688
                                                            CIP
                                                             AC

Facts On File books are available at special discounts when
purchased in bulk quantities for businesses, associations,
institutions or sales promotion. Please contact the Special
Sales Department of our New York office at 212/683-2244
(dial 800/322-8755 except in NY, AK or HI).

Designed and produced by BLA Publishing Limited,
East Grinstead, Sussex, England.

A member of the **Ling Kee Group**
LONDON · HONG KONG · TAIPEI · SINGAPORE · NEW YORK

Phototypeset in Britain by BLA Publishing/Composing Operations
Color origination in Hong Kong
Printed and bound in Portugal

10 9 8 7 6 5 4 3 2 1

Note to the reader
In this book some words are printed in **bold** type.
These words are listed in the glossary on page 42.
The glossary gives a brief explanation of words which
may be new to you.

Contents

The Moon's beginning	6	Mare Orientale, the Eastern Sea	26
Hot early days	8	Lunar highlands	28
The cooling crust	10	Inside the Moon	30
An asteroid hits the Moon	12	Exploring the surface	32
Creating a lunar "sea"	14	Shadows, rocks, and hills	34
The Imbrium Basin	16	Missions to the Moon	36
The impact process	18	Man on the Moon	38
Lava flooding	20	Apollo 17, the last visit	40
More lava, more craters	22	Glossary	42
Dating lunar surfaces	24	Index	44

Foreword

Long before man had the ability to explore space, he imagined what it might be like. Now that our journey beyond the Earth's atmosphere has begun, an artist can draw upon scientific knowledge gathered by probes and satellites in space, and observations made here on Earth, to portray a world where man has never been. Using what scientists know about the violent beginning of our solar system, the artist can take us back into the past to witness the formation of the planets, or into the future to imagine how they might one day be colonized. Through these paintings we can dive into the clouds of Jupiter, hover above the furnace of a sunspot, or even look back on our own solar system as we travel farther away into the galaxy.

In the six volumes of the **Planetary Exploration** series, we have combined the most advanced knowledge about the planets of our solar system with the extraordinary work of a noted space artist. Each book, written by an expert in the field, takes the reader beyond current facts and theories to the frontier of the unknown: the surface of Mars, the rings of Saturn, the tiny glacial world of Pluto, the many moons of Uranus, and beyond. Artist Don Davis has matched these exciting scientific discoveries with vivid illustrations that allow us to "travel" to these planets and unlock their mysteries.

We hope that, in **Planetary Exploration**, you will enjoy sharing this adventure.

David W. Hughes

The Moon's beginning

Planet Earth has a satellite which we have named Moon. Its **mass** is 1.2 percent of the Earth's mass and it is slowly moving away from us at about 1 in (3 cm) every year. Unfortunately we are still not sure where it came from. We do know that it is lifeless, dry, and dusty, but this has not always been the case.

How the Moon developed

The Moon formed from a **nebula** of gas and dust swirling around the Sun. As this cooled, particles of rock and iron condensed out, like rain drops forming in a cloud. Eventually, many cold rocky bodies were orbiting close to the Sun. Slowly, the bigger ones grew by hitting and absorbing the smaller ones. Soon the stronger **gravitational fields** of the larger bodies caused the smaller bodies to hit their surfaces faster and faster. The energy of these fast-moving, cold objects was converted into heat as a result of the collisions in a process called **accretion**. Big bodies, like the early Moon and the inner planets, grew by accreting smaller ones. The Moon grew hotter until it was a molten mass of red-hot, glowing rock and iron. It then became spherical and the heavier iron sank to the middle, forming a small core.

Facts about the Moon	
Diameter	2,160 miles (3,476 km)
Mass	1/81 of the Earth's
Distance from Earth	closest 221,467 miles (356,400 km) furthest 252,723 miles (406,700 km)
Average distance	238,900 miles (384,400 km)
Period of rotation	27 days 7h 42min
Orbital period	27 days 7h 42min
Inclination of equator	6.7° to orbit
Inclination of orbit	5.1° to Earth's orbit
Surface gravity	1/6 of the Earth's
Temperature	day 225°F (110°C) night −274°F (−170°C)
Atmosphere	none

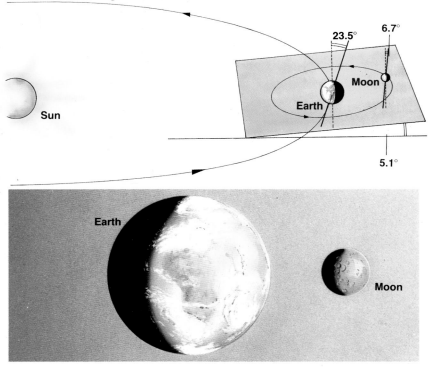

◀ The Earth is shown here in orbit around the Sun, and the Moon in orbit around the Earth (not to scale). The distance between the Moon and the Earth is only 0.26 percent of the distance between the Earth and the Sun. The angle between the orbits of Earth and Moon is 5.1° and the angle between the Moon's orbit and its equator is 6.7°.

◀ The Earth and Moon are drawn to scale. Earth is 3.7 times bigger, 81 times more massive and 1.7 times as dense as the Moon.

▶ The Moon was "born" just over 4,600 million years ago at a time when the dust in the cooling solar system nebula had accumulated to form big rocks. These came together to form even larger bodies. As the Moon "grew," a lot of material was lost. As it got hotter, many of the gassy products in the rocks escaped into space.

▶ The Moon grew slowly. We are not sure just how long the process took, but estimate it to be 100,000 to 1,000,000 years. As the Moon grew, so did its gravitational field. This enabled it to draw in material from greater and greater distances making the impact collisions stronger. Much heat was generated and **radiated** into space, but there was still enough to warm up the Moon until it became a spherical mass of molten rock and metal. At that time, it was spinning around on its axis almost every eight hours.

▶ At this stage the barrage of material falling into the molten surface is at its greatest. Each rock releases enormous amounts of energy on impact and turns the surface into a churning sea of white-hot, molten stone. Rocks hitting the surface cause a multitude of splatters and ripples and gas bubbles off, escaping because it is too hot to be held by the Moon's weak gravitational field. The new Sun is extremely active at this time too. Material from the Sun is rushing outwards in "gales" through the solar system sweeping away all trace of the tenuous atmospheres of the newly-formed planets.

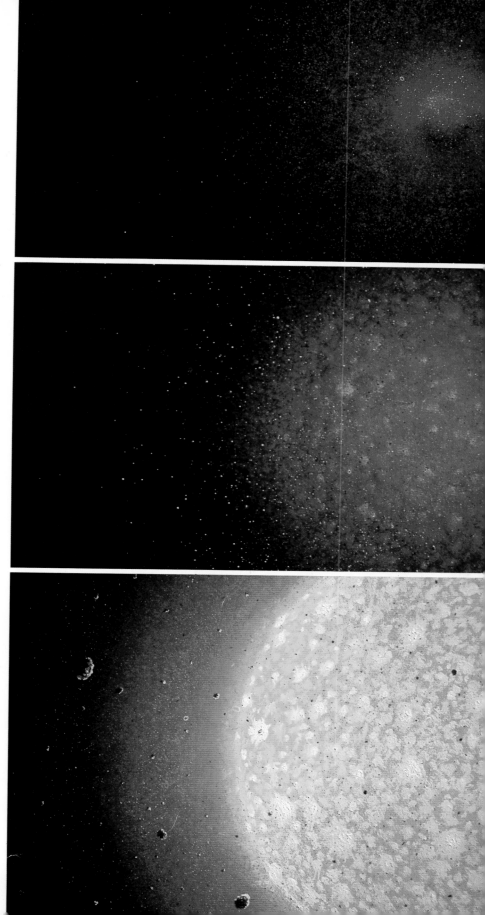

Hot early days

Rocky bodies, even today, continue to rain down on the Moon in a whole range of sizes. For every one that is around 60 miles (100 km) across, there are about 100 that are 6 miles (10 km) across; 10,000 that are 0.6 miles (1 km) across; 1,000,000 that are 110 yds (100 m) across, and many more even smaller. The first rocks had a composition similar to those that formed the Earth. Like Earth rocks, they contained considerable amounts of **water of crystallization**, carbon and oxygen. The rocks broke up on impact and considerable energy was released, which heated them to melting point. During this process they emitted large amounts of steam, carbon monoxide, and carbon dioxide, which formed a temporary atmosphere.

Why was the atmosphere temporary?

Two things affect the ability of a planetary body to hold an atmosphere. The first is its gravitational field. The larger this is, the stronger is the attraction that it exerts on the gas molecules. The second is the temperature of the gas. Molecules in a hot gas move faster than those in a cold one. Think of throwing a ball into the air. The harder you throw, the faster the ball travels and the higher it goes. If you were on the Moon and could throw the ball hard enough to reach a speed of 1.5 miles a second (2.4 km a second), the **escape velocity**, the ball would never come back. If the molecules in the Moon's atmosphere had that velocity, they could escape very quickly too.

▶ One hundred million years have passed since the picture below. The Moon has formed a crust on its surface and the heavy metals have sunk towards the center. The rate of rock bombardment has dropped by one-third.

Cooler regions of the crust show the overlapping scars produced by rock impact. The white circles around recent impact scars are produced by shock waves. These move outwards, like ripples on a lake after you have thrown in a pebble. The shock wave regions are slightly cooler than the surrounding areas.

4,500 million years ago ▶

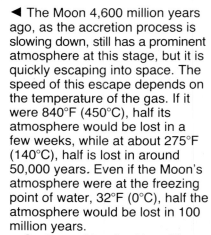

◀ The Moon 4,600 million years ago, as the accretion process is slowing down, still has a prominent atmosphere at this stage, but it is quickly escaping into space. The speed of this escape depends on the temperature of the gas. If it were 840°F (450°C), half its atmosphere would be lost in a few weeks, while at about 275°F (140°C), half is lost in around 50,000 years. Even if the Moon's atmosphere were at the freezing point of water, 32°F (0°C), half the atmosphere would be lost in 100 million years.

Cooling red-hot "rockbergs" can be seen floating on the otherwise molten surface of the Moon.

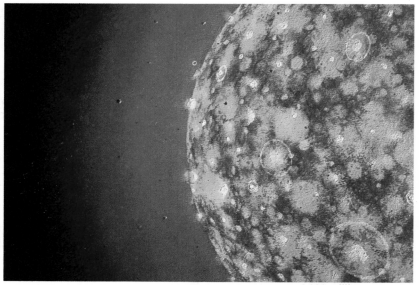

4,600 million years ago

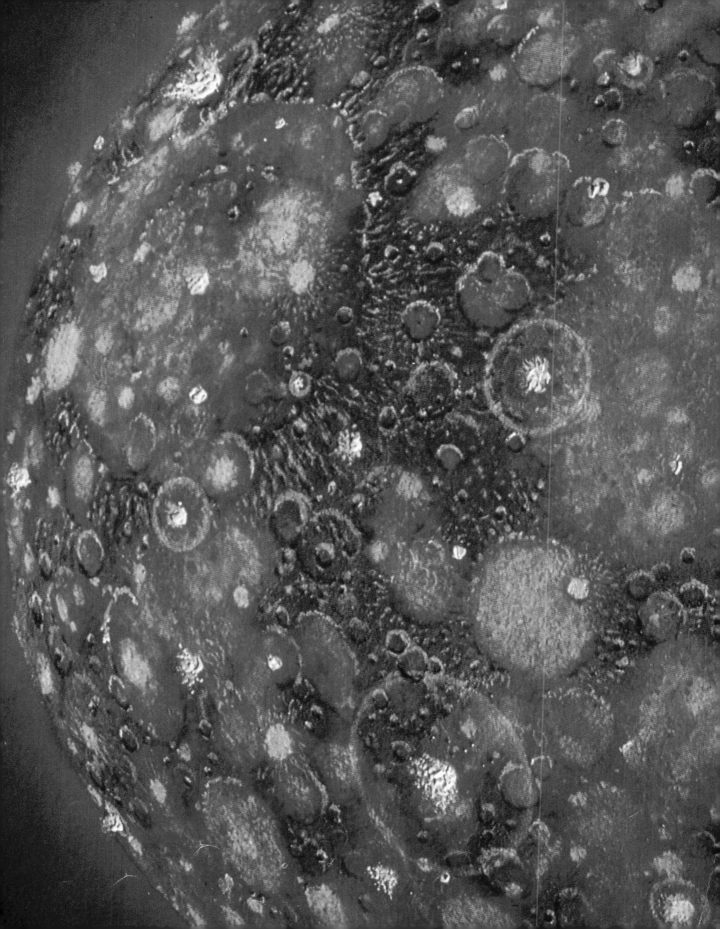

The cooling crust

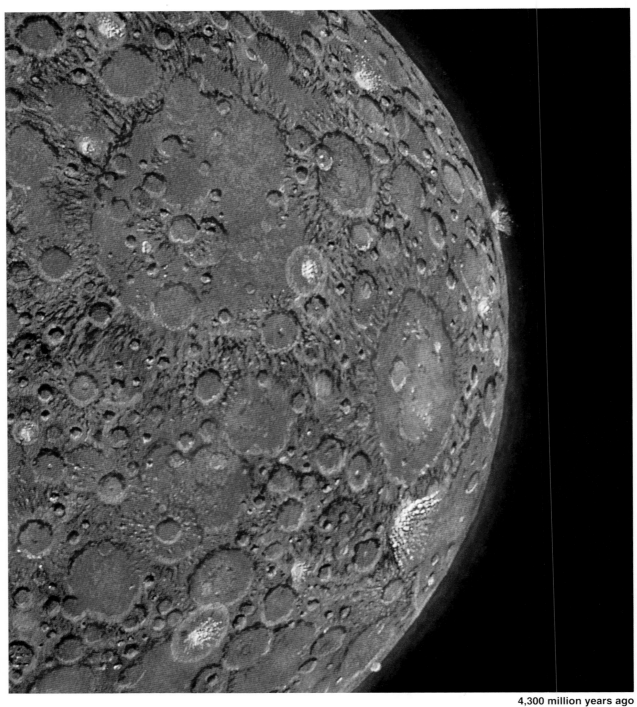

4,300 million years ago

◀ The Moon's atmosphere has escaped. The impact rate is one-twelfth of the rate in the previous picture. The growth by accretion of the Moon and the inner planets has all but ceased. Thick layers of the crust have cooled, and only an occasional rock impact gouges out a huge circular hole.

▶ At this time the Moon's surface is much cooler. It has been altered many times by continuous rock impact. Craters are still being formed here, but they simply replace the craters which were there before. The impact of large rocks has excavated vast depressions and the Moon's surface appears much rougher than it is today. These depressions will be flooded later by lava to form the maria, or "seas."

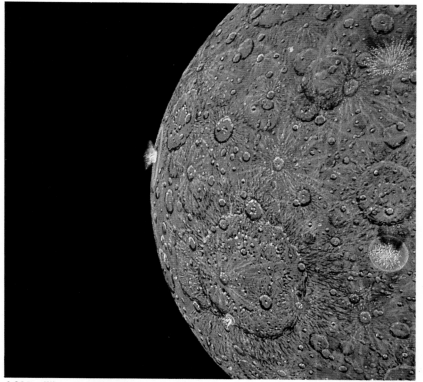

4,000 million years ago

▶ Another 300 million years have passed and the Moon's surface is now cold. The scene is set for the giant impact that resulted in Mare Imbrium, the huge circular "sea" which is near the north of the Earth-facing side of the Moon. The overall appearance is very similar to that of the present day far side of the Moon.

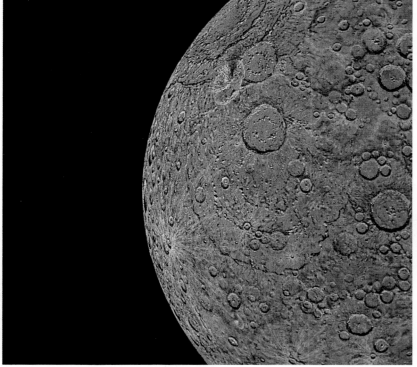

3,700 million years ago

An asteroid hits the Moon

The Moon was affected greatly by **tides**. The Earth was much closer to the Moon then than it is today. The tides slowed down the lunar "spin" and the Moon became tidally "locked" to Earth. So now the Moon takes as long to spin around once on its axis as it does to travel once around its earth-centered orbit. This locking makes one face of the Moon always point towards the Earth. The other face points away, towards the star-filled sky.

The exact origin of the Moon is still a mystery. This is frustrating to astronomers because one of the main goals of the US lunar exploration program was to answer this riddle. Between July 1969 and December 1972 twelve men walked on the Moon's surface. They collected and brought back to Earth 840 lb (381 kg) of rocks. Unfortunately it seems that none of this rock was the original **genesis rock**, for which the astronauts had been trained to look. The genesis rock seems to have been pulverized by the massive bombardment that took place in the early days of the Moon's existence.

▼ The enormous explosion which formed Mare Imbrium was probably caused by an asteroid 22 miles (35 km) wide hitting the Moon. At the point of impact it would have been traveling at 9 miles a second (14 km a second). This is a speed of 31,000 miles an hour (50,000 km an hour). Clouds of dust and gas can be seen moving away from the impact site. A shock wave, traveling faster than the ejected material, is pulverizing the surrounding lunar crust. The asteroid disintegrated on impact.

▲ The interior of the Moon is still hot and tar-like. Red-hot debris from the impact area is thrown upwards and outwards, scooping out a hole shaped like a cone. The top is several hundreds of miles across but is only temporary. The lunar crust below this vast cone and the crust that extends far beyond its rim, is **liquefied** by the gigantic shock waves that reverberate throughout the region.

Origins of the Moon

Today there are five main alternative ideas about the origin of the Moon. The first is that the Moon formed somewhere else in the solar system, and was captured by the Earth at a later date. This could explain why the Moon's composition is slightly different from the Earth's. It has less iron and less of the materials that **volatilize** easily. In the second theory, the Moon and the Earth formed at the same time. At this early stage in its evolution, the Earth had a system of rings, similar to those orbiting Saturn today. The particles in those rings came together to form the Moon. The third theory is that the Earth split into two very early in its life. As Earth was then spinning around every four hours or so, the effect of the Sun on Earth's tides would have produced a "tidal bulge," making the Earth pear-shaped. This "pear" then would have become progressively longer until eventually the tip would have broken off to form the Moon.

According to the fourth idea, the Earth was struck a glancing blow by a very large asteroid. The material that makes up the Moon was "splashed off" by the impact. The final idea is that a loosely-bound planet passed close enough to Earth to be broken up by forces exerted by the tides. The Moon was formed out of some of the bits of this passing planet.

These are the five ideas, but it is up to you, the scientists of the future, to find out which, if any, is correct.

Creating a lunar "sea"

◀ The formation of the vast Mare Imbrium is near its climax and the diameter of the crater is close to its final measurement of 746 miles (1,200 km), 12 times wider than the asteroid that initially hit the Moon. A tremendous amount of material is thrown out of the crater in a near-molten form. Material around the rim of the original cone is sliding into the excavated cavity. The material underlying the center of the crater that was pushed down by the initial explosion now bounces back. In smaller craters, this would lead to the production of a mountain peak in the center. Red-hot balls of rock fall back toward the surface. These will produce other small secondary craters.

▶ The expanding ring of rippling rock reaches its maximum width. The inner basin is solidifying and the frothy material hardens. The basin is surrounded by a huge ring of settling dust and a splattered and scoured surface is revealed. The higher the energy of the **ejector**, the further it travels. These molten "bombs" of rock tumble as they fly. Some break apart in midflight. When the fragments hit the surface they produce a chain of secondary craters. Some of these can be seen at the foot of the picture, aligned with the center of the Imbrium Basin.

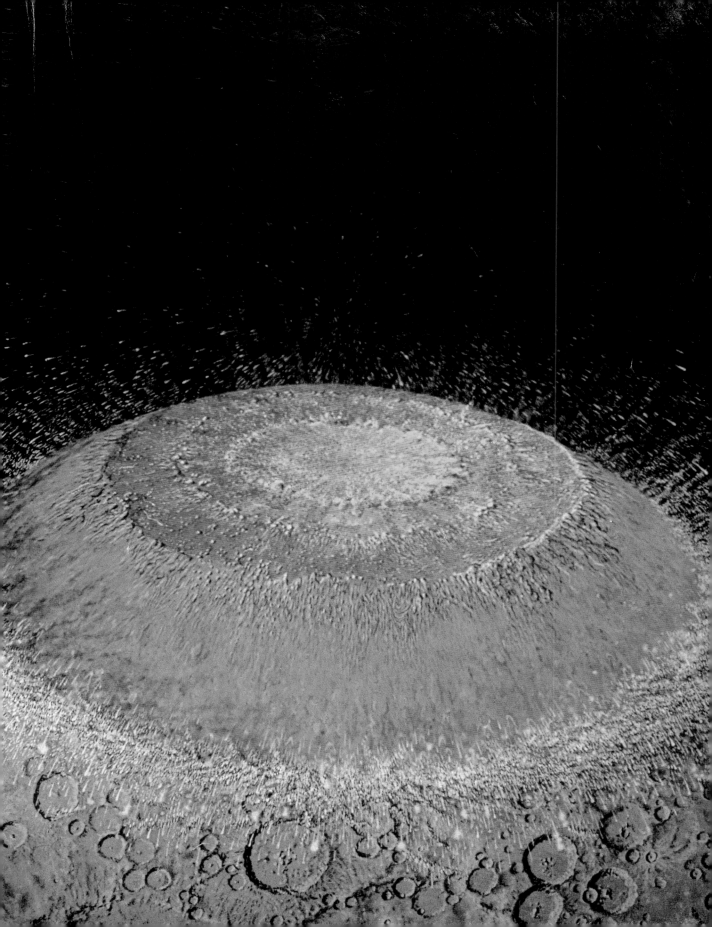

The Imbrium Basin

A complicated relationship exists between the size of a crater on the Moon and the size of the asteroid which created it. Mare Imbrium could have been produced by an asteroid 60 miles (100 km) across hitting the surface at a velocity of 1.9 miles a second (3 km a second), or one 22 miles (35 km) across moving at 8.7 miles a second (14 km a second), or one 7.5 miles (12 km) across moving at 43.5 miles a second (70 km a second).

We have no idea which event might have caused the crater, it all happened so long ago.

Nothing hits the Moon at a speed greater than 43.5 miles a second (70 km a second) or it would have enough energy to escape from the solar system altogether. Most of the asteroids which hit the Moon would have been traveling around the Sun in the same direction as the Earth. Instead of having a head-on collision, the objects would have collided while moving in the same direction along the same path, but at different speeds. The collision velocity, therefore, would be relatively low, somewhere in the range of 1.9 to 9.3 miles a second (3 to 15 km a second). This is 7,000 to 34,000 miles an hour (11,200 to 54,700 km an hour)!

▼ The dust has settled and the Imbrium Basin still glows, but is cooling. The rebounding of the ripples of molten rock has led to the formation of three rings. During the solidification of these rings, other minor rock movements have produced **concentric** and **radial** fractures. The cloud of ejector is shaped like a donut and is still expanding.

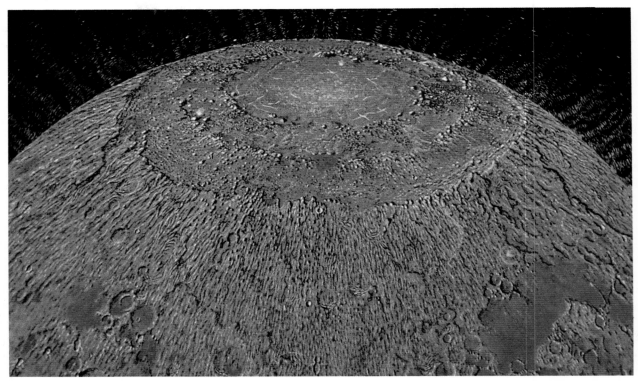

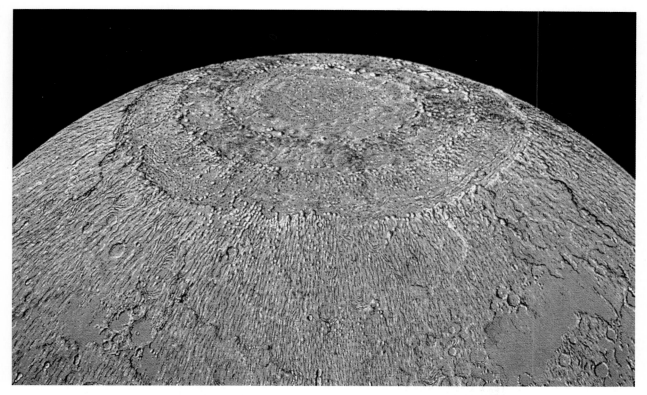

▲ Three-quarters of an hour has passed since the asteroid struck. The sky is clear. The material that was blasted to escape velocity has left the Moon far behind. The remaining debris has been scattered over more than half of the Moon's surface. The end product is the formation of a major impact basin, which is surrounded by mountainous rings of rock. The height of these rings decreases as the Moon adjusts to its new shape and the redistribution which has occurred in its mass.

The Moon's craters

One of the most interesting findings of the last thirty years of space exploration is that all the rocky and icy surfaces of the planets and satellites in the inner solar system are covered with craters. Mercury, the planet which is closest to the Sun, is pockmarked. So is Iapetus, the outermost of Saturn's larger satellites, a body which is about 10 times further away from the Sun than is the Earth.

Impacting rocks have molded many of these surfaces. Even planet Earth has some craters, but it is clear that we do not have as many as the Moon. The number of asteroids hitting a given area of Earth in, for example, a million years is estimated to be nearly the same as the number hitting a similar area of the Moon over the same time, but the Moon has no atmosphere. There are no dust storms to grind down the craters, and no oceans nor rain to wear them away. In addition, we know that the Earth's continents drift about, covering up some regions of the surface and revealing new ones, while the Moon has no continents. All these processes wiped out the vast majority of the record of crater formation here on Earth.

Gravitational fields affect how far the ejector travels; the lower the field, the further the debris can be thrown, so it is scattered more widely over the Moon's surface, than the Earth's — another reason for the difference in appearance.

The impact process

An asteroid moves so quickly that it has considerably more energy than the same amount of TNT explosive. The asteroid responsible for Imbrium had an energy equivalent to 2,000 million million tons of TNT.

How craters form

Craters form in three stages. First, the asteroid penetrates the surface like a bullet. The shock wave produced by the asteroid slowing down compresses the underlying material and the asteroid is broken up and vaporized. A "light flash" is produced and material jets away at high velocity from the impact.

Second, the compressed material relaxes, and the shock wave bounces back. More waves expand outwards from the point of impact, leading to the ejection of a large mass of material at low velocity and very shallow angles from the rim of the new crater. As the crater deepens more is thrown out at steep angles. Some of this material will then fall back and cover the new crater with a blanket of loose debris.

The final process is called modification. The walls slump and produce a series of terraces around the rim.

The central peaks contain tilted, jumbled rocks, which were thrown up by the rebounding shock wave. The enormous energy released can melt the warm rocks below the lunar surface and this molten material then squeezes out through cracks in the crust.

▶ One hundred million years have passed since the formation of the Imbrium Basin. The Basin is largely undamaged but some smaller craters have been formed inside it. Also lava from inside the Moon has welled up through cracks under the crater, spreading over the surface and filling some of the low-lying parts. When another large basin, called Orientale, was formed on the far side of the Moon, the impact produced strings of secondary craters. These can be seen at the lower left.

3,600 million years ago ▷

▶ This is a view of the formation of the Imbrium Basin, as it would have appeared from Earth. Earth, in contrast to the airless Moon, has a considerable atmosphere fed by volcanic gases. The craters on Earth are filled with water.

◀ The crater is formed by the **hypervelocity** impact of rock, where the incoming asteroid is moving faster than the speed of sound in the rock. The crater caused by this impact is much larger than the asteroid which created it. Many of the underlying layers of the lunar crust have been turned over at the crater's rim. Ejected debris from the crater has spread over a large circular area, and much of it has fallen back into the crater.

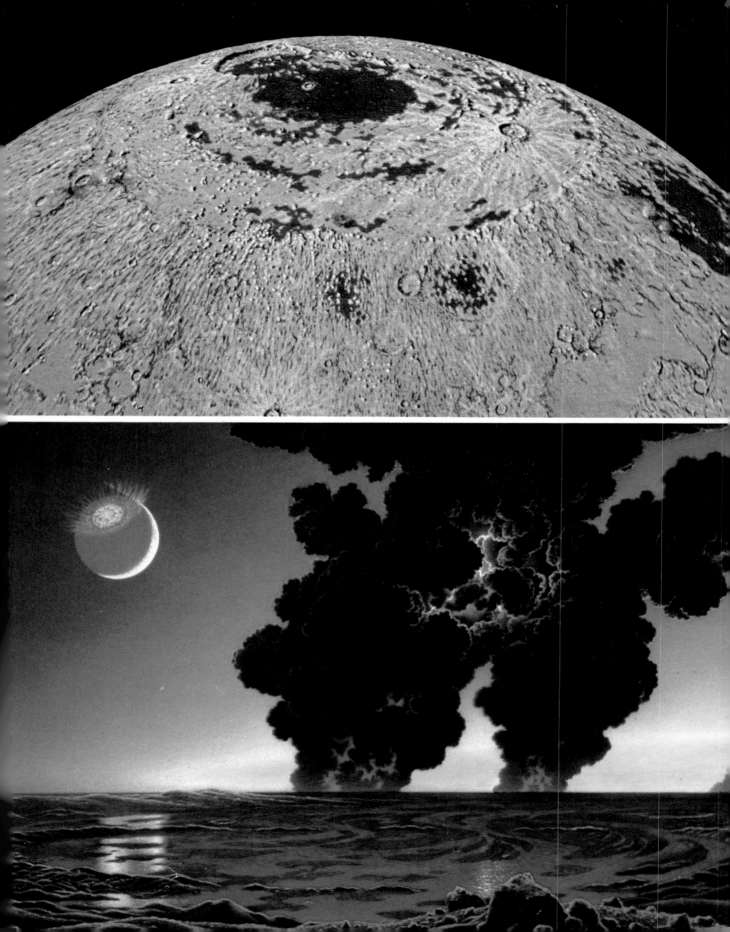

Lava flooding

The lunar maria are "seas" of volcanic lava, which have a similar composition to the **basalt** rocks still building up the volcanic islands of Hawaii and Iceland. About 17 percent of the lunar surface is covered with this infill, the large majority of it on the side of the Moon facing Earth. It is highly unlikely that all the mare basins were filled in one outpouring. The lava probably issued as a series of sheets, each being a few tens of yards (meters) thick and each solidifying before the next one flowed over the top. This resulted in the maria's tilted surfaces.

How the "seas" formed

Early in its life the Moon was covered with a layer of **low density** rock which was about 30 miles (50 km) thick. This is still visible today in many regions, especially in the southern highlands where a series of very deep holes was formed by impacting rocks, which also cracked the underlying lithosphere, the region between 30 and 625 miles (50 and 1,000 km) deep. **Magma** flowed through the cracks into the basins until the surface pressure was equal to the pressure at the bottom.

Then the lava cooled. As it solidified, it became denser and shrank, causing its surface to drop by about 1.25 miles (2 km). Where the cracks remained open, fresh lava flowed in and a 1.25-mile thick (2-km thick) layer of lava was added to the basin. The lava is denser than the surrounding rock so the maria region has a greater mass than the surrounding highland areas. This increases the gravitational field, producing what is known as a **mascon**—Mare Imbrium and Mare Serenitatis are the largest. These were discovered in the early 1960s because they affected the paths of the spacecraft that were put into orbit around the Moon.

The far side of the Moon

Nothing was known about the far side of the Moon until the flight of the Soviet spacecraft, Luna 3, in 1959. Unlike the face that points towards Earth, the far side has very few mare. It is made up of light-colored, rugged, highly cratered mountains just like the southern highland regions on the near side. It also has large basins which have not been filled with lava, although we are not sure why. The far side of the Moon has been pointing away from the Earth for a very large part of the Moon's life. This seems to have affected the lunar interior. The crust on the Earth side is only 37 miles (60 km) thick, but it is 60 miles (100 km) thick on the far side. Also the Moon is slightly egg-shaped, with the small end pointing towards Earth.

▶ Eight hundred million years have passed. Much of the Earth-facing side of the **Moon** has been inundated with lava. The Imbrium Basin has completely filled up with lava because the initial impact occurred in a region where the surface was already depressed. In places, the lava has breached the rim of the crater.

Only the very highest ring of mountains remains visible. The speed with which this flooding occurred is indicated by the smoothness of the solidified lava. No new craters have formed in it yet.

2,400 million years ago ▷

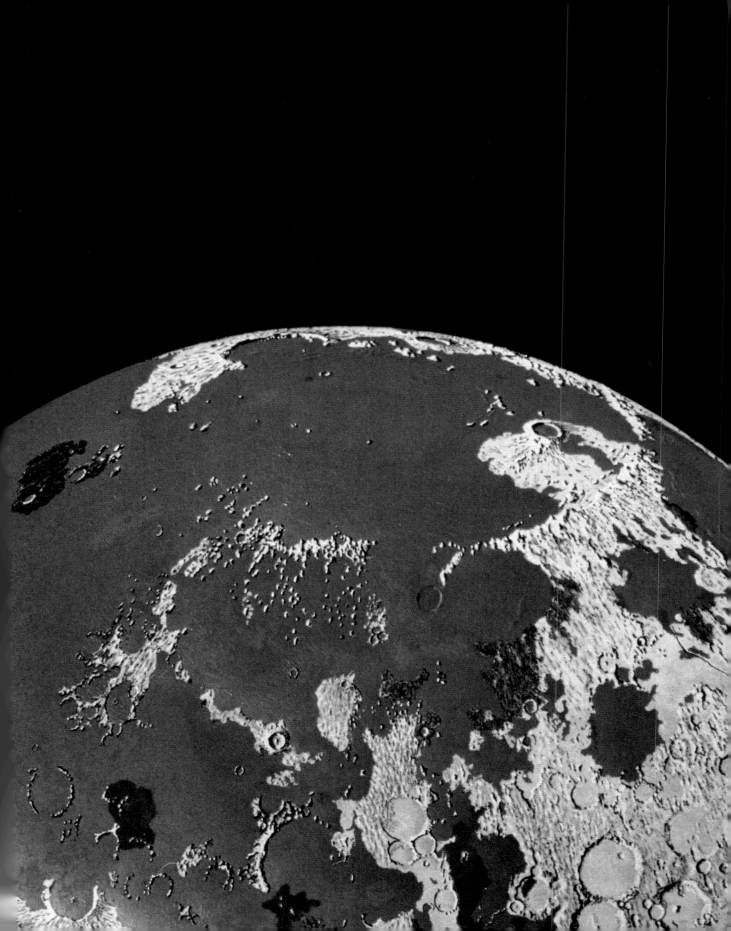

More lava, more craters

The large pools of lava spreading out over the basins extend for up to 600 miles (around 1,000 km) and can flow over very shallow slopes. The lava gushes out quickly and has a low **viscosity**. As it cools, the lava shrinks, leaving a visible "high tide" mark around the edge of the basin. The surfaces of the maria are rough and are marked by a variety of features. Sinuous **rills** cluster around their edges, usually a few miles (kilometers) wide and sometimes hundreds of miles (kilometers) long. They are similar to the Earth's rift valleys between two faults in the crust. The rills cut across both craters and mare. "Wrinkle ridges" protrude from the lava plains forming concentric rings in the mare.

Most scientists think the rills were produced when the underlying layers of lava folded as they cooled and shrank. In fact, they look just like the wrinkles in a rug which has been pushed against a wall. Dome-like features with craters at their summits, like Earth's volcanoes, can also be found.

The crater Eratostenes

A prominent crater, Eratostenes, is surrounded by lighter-colored material, which has been ploughed up by ejected bombs of rock from the initial crater. The light-colored rocks are not chemically different from the surface, but their brightness is due to the soil being churned up and full of holes. As time passes, many small cosmic particles will hit the area, forming tiny craters and compressing the soil, making it reflect less light and look darker. Lunar soil only reflects about 5 percent of the light that falls on it. Exposure to radiation also makes the soil blacker.

Notice that Eratostenes has a central mountain peak. The diameter of the crater is about 37 miles (60 km) and it is around 1.8 miles (3 km) deep. The largest craters on the Moon are only about 3 miles (5 km) deep, shaped like shallow saucers. The lunar landscape is gently undulating and not as dramatic as it looks during certain phases of the Moon. Smaller craters are deeper and bowl-shaped. Their depths are about one-quarter of their diameters.

The shape of craters

The majority of the craters on the Moon are round. This may seem unusual when we remember that the asteroids which produced the craters hit the surface at any angle. When the asteroid buries itself into the surface, the explosion is enormous and the crater which is formed is much bigger than both the asteroid which caused it and its path through the crust. The

▶ Five hundred million years have passed since the previous picture. The rate of crater formation has dropped again and only a few new craters have been formed. One of the most prominent is Eratostenes, which is on the lower rim of Mare Imbrium (near the center right). Long, light-colored strips of ejector radiate out from these craters. One of the last of the major lava flows can be seen spreading out over the left side of Mare Imbrium, adding another layer of material to the basin.

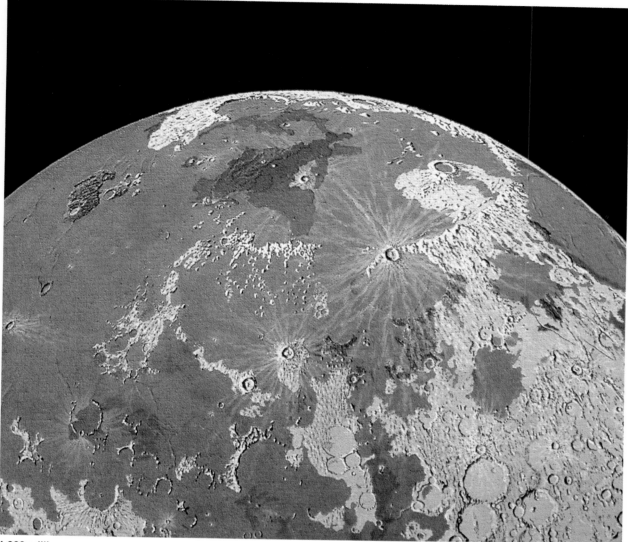

1,900 million years ago

shape of the crater does not depend on the direction from which the asteroid came. Only if an asteroid "grazes" the surface is an elongated crater formed. The craters Schiller and the Messier twins are perfect examples of this type.

Lunar features were first named in the 1600s. The dark areas were thought to be expanses of water so they were given Latin titles like Oceanus Procellarum (ocean of tempests), Mare Imbrium (sea of showers), Lacus (lake) and Sinus (bay). An Italian Jesuit priest, Giovanni Riccioli, published a map of the Moon in 1651 and named the mountains after ranges on Earth such as the Alps and the Apennines. The craters were named after famous people who were usually, although not always, astronomers. Riccioli's system has survived to this day and has been used to name features on the far side of the Moon.

Dating lunar surfaces

On the surface of the Moon there are many more small craters than large ones. The number of craters also varies greatly between areas. There are considerably fewer craters in the mare regions than in the highlands. The reason is simple. Where an area of the Moon's surface has been bombarded over time by asteroids and meteorites (small fragments that break off asteroids when they collide) more craters will form. Lava flows wipe the slate clean. So regions with many craters are older than regions where the craters are sparsely distributed.

Aging craters

To measure the age of certain regions, we count the number of craters of a particular size that appear. This number is then compared with the rate at which asteroids and meteorites hit the Moon in general. For example, over a typical region of the Moon with an area of 390,000 square miles (1 million square km) there would be about five craters with diameters larger than 60 miles (100 km), 500 larger than 6 miles (10 km), 50,000 larger than 0.6 miles (1 km), and so on. One problem with this method is that the highland regions have suffered greatly from "saturation bombardment." Here, the newly-formed craters obliterate, erode and bury the smaller ones.

To overcome this difficulty we count only the larger craters. The results obtained by this method can now be compared with results obtained from analyzing the samples of Moon rock which were brought back by the Apollo astronauts. One of the methods relies on the **radioactive decay** of atoms. For example, radioactive potassium 40 decays into an **inert gas**, argon 40. In 1,300 million years half of the potassium 40 has decayed. In another 1,300 million years half of the remaining potassium has decayed, leaving only one-quarter of the original amount. This process continues, with the potassium being reduced by half every 1,300 million years.

If we can measure the proportion of potassium 40 to argon 40 in a rock, we can then find out how long the rock has been in a solid form, so preventing the argon from escaping.

The time a specific lunar rock has been on the surface can also be measured. When it is first exposed, it is hit by energetic atoms coming from the Sun, known as "cosmic rays," which break up the atoms in the surface skin of the rock. The amount of damage done can be measured and the result tells scientists how long the rock has been on the surface.

▶ The northern polar region of the Moon looks in this painting just like it does today. The prominent crater at the center of the picture is Copernicus, formed about 900 million years ago. The crater to the left of Copernicus is Kepler and to the left above that is Aristarchus. Both are around 500 million years old, Aristarchus being slightly the younger.

These new craters have no lava in them. Older ones, such as Plato (at the top) and Archimedes (inside Mare Imbrium and on the right) are both flooded with lava. At the bottom left, in the dark regions of Oceanus Procellarum, is a perfect example of a **ghost crater** (named Flamsteed). Here, only the topmost peaks of the mountainous ring can be seen as they peep through the lava.

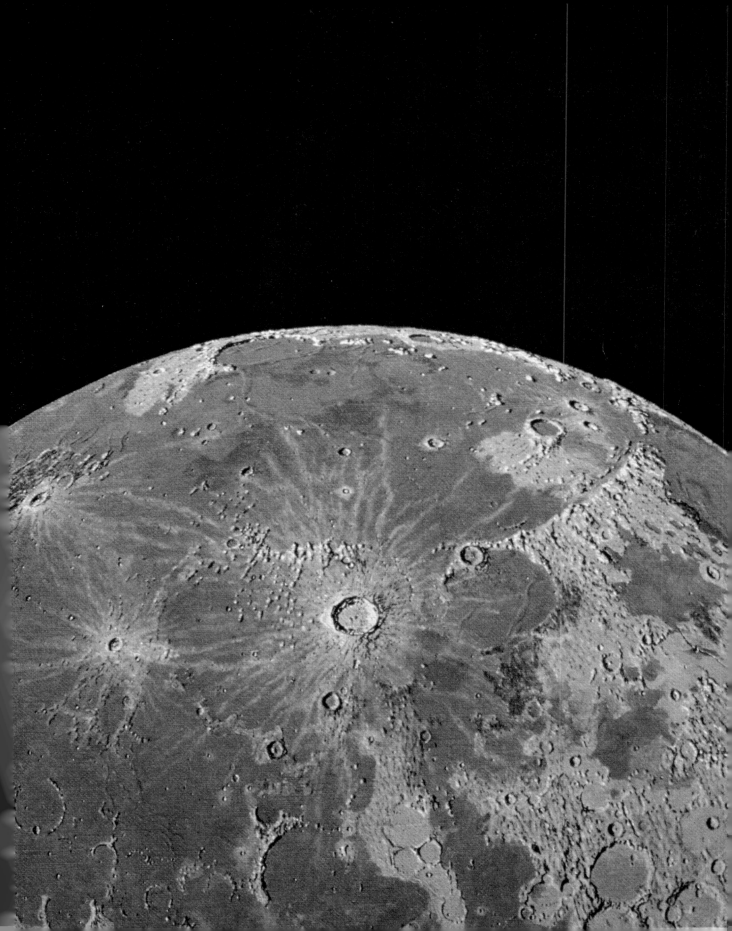

Mare Orientale, the Eastern Sea

◀ The large, 560-mile (900-km) diameter, impact scar in the center of this picture is Mare Orientale, the Eastern Sea. The dark lava plane of Oceanus Procellarum is at the top left, while just below Procellarum is the crater Grimaldi, the inner ring of which has been flooded by lava.

The Orientale Basin is the Moon's most spectacular and best-preserved example of a giant multiringed impact basin. It was formed 3,850 million years ago and is on the very eastern edge of the Earth-facing side of the Moon.

At least three concentric rings of mountains can be seen. There is a large radial gouge stretching towards the south, and an extensive blanket of ejector around the basin has almost blotted out pre-existing craters.

The center of Orientale is a mascon, made up of molten ejector which rained back down onto the surface. It has been estimated that the total volume of molten material produced by the formation of the basin measured more than 48,000 cubic miles (200,000 cubic km).

The outer mountains

The outer ring of mountains is about 3.7 miles (6 km) high. The three rings around the basin are caused by gravity forcing the surrounding rock to settle during and after the formation of the craters. To explain this, think of making a hole in a substance like water, a substance which does not keep its shape. It would fill in immediately and the surface would become level again. If a crater in the lunar crust is very large and deep, the stresses in the rock underneath the crater are greater than the strength of the rock. Rings of the rock then break up and slide down into the center, producing a mound. From above, the edges of the slipped rings form circular mountain ranges.

Crater formation today

By carefully measuring the orbits of the asteroids which have paths crossing the orbit of the Moon, it has been found that a new 0.6-mile (1-km) diameter crater will be formed every 50,000 years, a 3-mile (5-km) crater every 800,000 years, a 6-mile (10-km) crater every 3 million years and a 31-mile (50-km) crater every 50 million years. This means that the face of the Moon has changed very little since it was first observed from Earth, so we would not expect to find any eyewitness records of crater formation.

Even so, on July 18, 1178, five men from Canterbury, England, were looking at the **new Moon** and saw that "suddenly the upper horn split in two. From the midpoint of this division a flaming torch sprang up, spewing out over a considerable distance, fire, hot coals and sparks. Meanwhile the body of the Moon which was below writhed, as it were in anxiety." This is thought to be an account of the formation of the crater Giordano Bruno.

Lunar highlands

The lunar highlands are the oldest exposed regions on the Moon, between 3,900 million and 4,000 million years old. They are richer in alumina than the maria rocks, and are saturated with large craters. The crater formation process has stirred and mixed the underlying material to a depth of 6 miles (10 km). It seems probable that the highlands are made up of material from the crust which was produced when the Moon was molten. At that time the Moon **differentiated**, the heavy metal elements like iron and nickel sank to the center and the lighter rock waste formed a mantle of rock around the outside.

The Moon's surface

The Moon is covered with a soil, called regolith, a blanket of dust and rock fragments, produced by meteorite bombardment. Meteorites reach the Moon's surface without being destroyed, while most heading toward Earth burn up in our atmosphere as "shooting stars." On the Moon, every meteorite produces a crater and breaks up more rock. The regolith is churned up and for the last 4,000 million years it has become thicker. Over the "new" maria, its average depth is about 15 ft (5 m). In the highlands it is between 33 and 99 ft (10 and 30 m) deep.

Before the space age, some scientists thought that the regolith was like quicksand and would swallow spacecraft that tried to land on it. Fortunately, this is not the case and people have walked on the Moon. Although it is fine-grained and loosely-packed, the Moon's soil is firm. It sticks together, like wet sand and astronauts left behind very clear footprints, which will remain for many millions of years, until being eroded by renewed meteorite bombardment. In places, the regolith is thin because landslides have swept it from the slopes into craters.

Hot days, freezing nights

The Earth's atmosphere tends to smooth out the extremes of temperature between night and day. This does not happen on the Moon where one day is 27.3 Earth days long. When the Sun is directly overhead, the temperature on the Moon's surface, near its equator, is about 225°F (110°C), hot enough to boil both water and blood! Fourteen days later at midnight the temperature has dropped to around −274°F (−170°C), only 86°F (30°C) higher than that needed to freeze air. The regolith, however, is a good insulator. At a depth of about 20 in (50 cm) below the surface the temperature is 4°F (−20°C). This hardly varies, only changing throughout the year by about 5°F (2.8°C).

▼ This illustration shows the western part of the Orientale Basin, where a profusion of mountains are strewn around the outer rim. The radial splash pattern merges into long chains of craters as one moves away from the center. The region on the left of the picture is typical of the ancient cratered highlands, which occupy 30 percent of the near side, and nearly all of the far side, of the Moon. These regions have virtually no lava. There is a small amount in the double-ringed basin, Apollo, in the lower left corner. In the upper left is the double-ringed basin, Hertzprung.

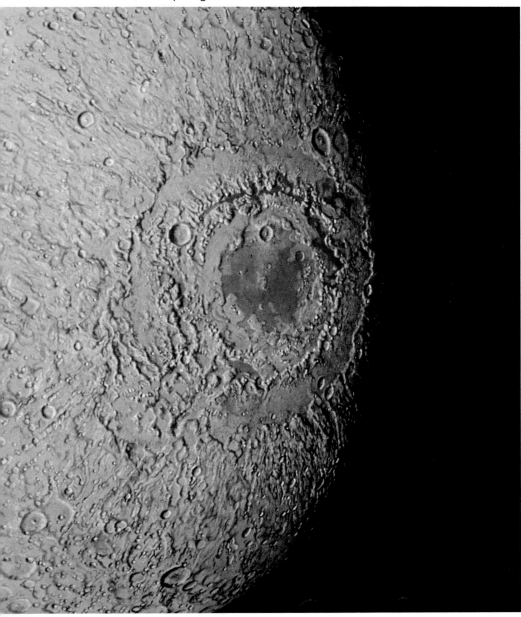

Inside the Moon

What color is the Moon?

At first sight the Moon appears to be gray and dull. More detailed studies show that the younger flows of lava are a blue-gray while the older ones tend to be a reddish-gray.

The light from the Moon we see on Earth is reflected light from the Sun. On average, the Moon reflects only 7 percent of the light which hits it, while the remaining 93 percent is absorbed. The ability to reflect light is determined by the roughness and fine-grained nature of the surface. The light hits the dusty soil and then bounces from grain to grain. Only a little light escapes, and this depends strongly on the angle at which the light first hits the surface. If the light struck it vertically, nearly 10 percent would be reflected. But if it came in at an angle of 20° to the vertical, only 5 percent would be reflected. At 60° to the vertical, only 2 percent is reflected. This phenomenon is known as the "phase effect" and accounts for the fact that the full Moon is very much brighter than the new Moon. When the Moon is full, sunlight strikes the surface vertically and a complete circle is seen.

▼ The first humans to see this view were the Apollo 8 crew as they returned from the far side of the Moon in December 1968. The sight of our small life-supporting planet, hovering some 238,900 miles (384,400 km) away, underlines both the thrill and danger of space exploration. The Moon's surface appears featureless and gently rolling because the Sun is nearly overhead. The blackness of the sky reveals the lack of atmosphere.

▲ The Moon's interior is colder than the Earth's, mainly because the ratio between surface area and mass is larger, so cooling is quicker. The molten core could be, at most, 400 miles (700 km) wide, and is surrounded by a mantle which, at great depths, could be partially molten. The crust is slightly thicker on the Earth-facing side.

The structure of the Moon

The Moon's average density is 3.3 times that of water and this helps us calculate its composition. Instruments placed on the surface by the Apollo astronauts have recorded thousands of "moonquakes," and "impactquakes" caused by meteorites or discarded rocket **stages** crashing into its surface. Analysis of these quakes enables us to calculate how the composition of the Moon varies with depth. The top 40 miles (60 km) is a crust of calcium- and aluminum-rich rocks. Beneath this is a layer of solid rock which is around 600 miles (1,000 km) deep, called the lithosphere. The outer part of the Moon is colder, thicker, and more rigid than the outer part of Earth, so the Moon has no volcanic activity or continental drift.

The region between 600 and 900 miles (1,000 and 1,400 km) deep, the asthenosphere, is the zone of partial melting. A small core composed of iron and sulfur may exist below 900 miles (1,400 km).

Earth's **magnetic field** is caused by **dynamo** currents in its molten core. No magnetic field is detectable on the Moon, but older rock samples do have a trace of magnetization. The Moon probably had a magnetic field in the past, which could have disappeared when the core solidified.

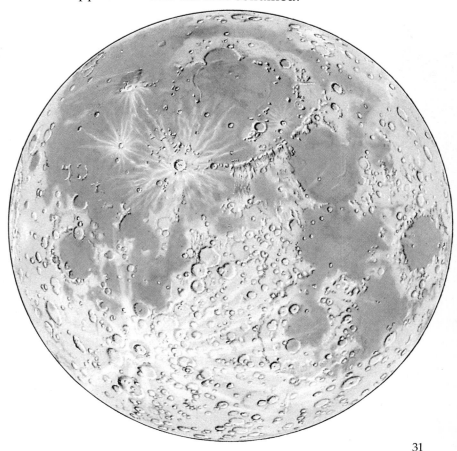

▶ This map has been drawn with the help of photographs of the Moon. The first of these was taken in 1840. The map shows the Earth-facing side. The division between highland regions and the seas and oceans of solidified lava can be seen easily.

Exploring the surface

Closer exploration of the Moon was one of the major aims of the space race, a race which started when the Soviet's Sputnik was launched on October 4, 1957. The United States responded with the "Kennedy challenge" when on May 25, 1961, President John F. Kennedy said, "I believe that this nation should commit itself to achieving the goal, before this decade is out, of landing a man on the Moon and returning him safely to Earth. No single space project in this period will be more impressive to mankind, or more important in the long-range exploration of space."

The problems in reaching this goal were immense. The Moon's characteristics were largely unknown. In fact, certain experts predicted that the surface would not support a person. Astronauts would sink down into a "quicksand" of dust, buried alive, disappearing like stones in a pond.

Much needed to be done. First, the surface of the Moon needed to be investigated in detail. The smallest object visible from Earth was 0.6 miles (1 km) across. Earth-based observation was clearly not detailed enough.

The Ranger probes

The Ranger probes were designed to crash land. Their job was to produce the first close-up pictures of the Moon. Television cameras on board took increasingly detailed pictures as the probes sped towards the surface.

In total, nine Rangers were launched, but only numbers 7, 8 and 9 (July 1964 to March 1965) were successful in their missions and together they sent back 17,000 photographs. Even the flattest part of the Moon was shown to be pockmarked with craters, as small as 3 feet in diameter. As the probes neared the surface, smaller and smaller craters could be seen, but these were the same shape as the larger ones. An astronaut could not

▼ The magnificent Copernicus crater lies to the south of Mare Imbrium and west of the center of the Earth-facing side of the Moon. It measures about 60 miles (100 km) wide and 1.9 miles (3 km) deep. The surrounding walls stand like cliffs 0.6 miles (1 km) above the adjacent ground. A group of mountainous peaks, in the center of the crater, form a range about 10 miles (16 km) long and 650 yds (600 m) high.

determine easily his height above the surface and would have to rely on radar.

The Lunar Orbiter

The Lunar Orbiter and Surveyor programs took place at about the same time. The Orbiter was designed to photograph the whole surface of the Moon in detail. Five craft were used: 1, 2 and 3 were put into orbits around the equator and 4 and 5 into orbits around the poles. They flew between August 1966 and August 1967. Each craft had two cameras on board, one for detailed shots and the other for wide-angle views. Their task was to search for landing sites. The photographs were recorded on 70mm film which was developed on board and then electronically "scanned," to convert the data to a form suitable for transmission back to Earth. When all the film had been used and read, the craft were crash landed onto the Moon so that they were not a potential hazard to future missions.

The Surveyor probes

The Surveyor probes paved the way for the manned Apollo craft. There were five successful Surveyors between May 1966 and January 1968. Each craft was shaped like a pyramid and had three footpads made from crushable aluminum. Each had a 9,920lb (4,500kg) **thrust** rocket, which slowed it down as it approached the Moon. The rocket was then discarded and three small jets took over the task until the craft was only 14 feet (4.3m) above the surface. Surveyor then dropped to the ground making impact at about 3 miles an hour (5km an hour). The craft had camera systems on board and the pictures taken were later made into fascinating lunar panoramas.

Sampling scoops were extended from the spacecraft and used to scrape up the soil, which was analyzed by bombarding it with the nuclei of helium atoms. The variations in energy of the scattered nuclei were examined. Surveyor 6 actually refired its rockets and mometarily lifted off from the surface, relanding some 2.7yds (2.5m) away. The Surveyor mission proved that the Moon was safe to land on.

From Earth we always look down into the crater.

These pages, and the next two, show a 360° panoramic view of the Copernicus crater as seen from a valley in the central mountains. In the view shown below, left, the Sun is directly behind us and about 30° above the horizon. The distant crater wall has a washed-out appearance.

Shadows, rocks, and hills

The Apollo program was an expensive and ambitious endeavor. Its aims were to take astronauts to the Moon, to land them so they could walk on the surface, and then return them to Earth. But, the Moon is 238,900 miles (384,400 km) away. There is no atmosphere to breathe. There is no water to drink. The temperature on the surface varies from 225°F to −274°F (110°C to −170°C) and back again throughout one month. To land successfully on the surface the exact spot and time have to be chosen carefully.

The Apollo missions

The Apollo spacecraft were launched from Cape Canaveral, Florida, on top of huge Saturn V rockets. There were 17 Apollo missions between 1966 and 1972. The testing of the ability of humans to live and work in space had been carried out using a series of close-to-Earth missions called Mercury and Gemini. These tests included activities outside the spacecraft, rendezvous and docking procedures with orbiting craft.

There were three main pieces of hardware making up Apollo. At the top was a Command Module, which was conical in shape, 12.9 ft (3.9 m) wide at the base and 10.5 ft (3.2 m) high, which contained three couches for the astronauts to lie on during takeoff and return to Earth. Behind this was a Service Module which supplied water, oxygen and power and housed the antenna for communication. This module also contained the propulsion system, rockets used for major speed changes.

At the top of the third stage of the Saturn V rocket, just under the Service Module, was the Lunar Module. This was used for landing on the Moon and for returning to the Command Module, which remained in orbit around the Moon.

At the start of the mission, Apollo was launched into an

▼ This is a continued view of the Copernicus crater. The walls in the distance are about 30 miles (50 km) away. They appear to be much closer. This is simply because there is no atmosphere to provide the absorbing haze that adds perspective on Earth. In the view, right, we face the Sun, so the backs of the nearby rocks are in shadow.

Earth **parking orbit**. After checking that all systems were working correctly, the final stage Saturn IVB rockets were fired, accelerating the Apollo Modules into a wider orbit to take them from Earth to the Moon. During this phase the Saturn IVB stage was jettisoned. On nearing the Moon, rockets were fired to slow down the craft enabling it to be captured into an orbit around the Moon. Two of the crew then crawled into the Lunar Module, detached it from the combined Command and Service Modules and descended to the Moon's surface.

The first man on the Moon

Neil Armstrong stepped onto the surface of the Moon on Monday, July 21, 1969. "That's one small step for a man — one giant leap for mankind," he said. He looked down at the fine, powdery soil. His boots sank in about 0.1 in (3 mm). The footprints in the surface could be seen clearly by scientists on Earth monitoring the mission. They were similar to footprints seen in damp sand. A few minutes later Edwin "Buzz" Aldrin walked down the ladder to join him.

Later, the Lunar Module blasted off from the Moon's surface to dock with the orbiting Command Module. Firing rockets then enabled the system to escape from the lunar gravitational field and move back towards Earth. On nearing Earth, the system decelerated, falling once again into an Earth orbit. Here the astronauts jettisoned the Service Module, re-entered the atmosphere of our planet, and finally splashed down in the Pacific Ocean. The journey from Earth to Moon took about five days round trip, with two and a half days spent on the Moon's surface.

In later missions Apollo astronauts visited Mare Tranquillitatis, Oceanus Procellarum, Frau Mauro, Hadley Rill, the Descartes Highlands and Mare Serenitatis. Apollo 15, 16 and 17 all used a lunar roving vehicle. A total of 840 lb (381 kg) of lunar rock was brought back to Earth and a host of experiments were placed on the Moon to monitor its activity and the lunar environment.

Missions to the Moon

Beginning with the Apollo 12 mission, an Apollo Lunar Surface Experiments Package (ALSEP) was set up at each landing site. Power for the experiments was provided by a **thermoelectric generator** heated by the radioactive decay of its uranium fuel. The instruments were designed to continue to work long after the astronauts had returned to Earth. In fact they worked for five years and over 9 million readings could be radioed back to Earth each day.

Scientific findings

A **seismic** experiment measured the vibrations of the lunar surface. In July 1972, a 220 lb (100 kg) meteorite hit the Moon. The vibrations that the impact set up made the Moon "ring" for over an hour. On Earth the vibrations would have ended sooner because the interior of Earth is hotter and, so, less solid. The number and energy of the electrically-charged particles blowing across the solar system from the Sun were measured. The traces of gas near the lunar surface were monitored. **Magnetometers** were used to see how lunar rock responded to changes in the interplanetary magnetic fields. One experiment was designed to measure the rate at which micrometeorites hit the Moon and the amount of moving **electrostatically**-charged dust near the lunar surface.

An array of **corner cubes** was set up at three of the Apollo sites. Pulses of laser light hitting this array bounced back along their incoming paths. By timing how long it took for the pulse to travel to the Moon and back, the Earth-Moon distance could be measured to the nearest 6 in (15 cm). This experiment can be continued indefinitely and will lead to a much greater understanding of how the Earth-Moon system is moving and evolving.

The most important scientific aspect of the Apollo program was the collection of lunar surface samples. With man present these could be chosen carefully. The position and orientation of each rock was documented accurately, to reveal important information about the impact history of the Moon. A rock found at one site could have been there for many millions of years or could have been blown there recently from a newly-formed crater. At one site the astronauts found some unusual orange soil, a sample of which was collected and brought back to Earth for analysis.

Moon Missions

September 1959, *Luna 2* (USSR): First spacecraft to crash land on the Moon.

October 1959, *Luna 3* (USSR) sends back the first photograph of the far side of the Moon.

July 1964 *Ranger 7* (USA) successfully takes 4,316 TV pictures as it flies towards its crash landing on the Moon.

January 1966 *Luna 9* (USSR) lands softly in Oceanus Procellarum and takes photographs for three days.

August 1966 *Lunar Orbiter 1* (USA) starts photographing the Moon from an equatorial orbit.

June 1968 *Surveyor 1* (USA) lands near crater Flamsteed.

December 1968 *Apollo 8* (USA), orbits the Moon 10 times.

July 1969 *Apollo 11* (USA) Armstrong and Aldrin walk on the Moon.

September 1970 *Luna 16* (USSR) First remote sampling mission brings back 3.5 oz (100 g).

November 1970 *Luna 17* (USSR) Automatic Moon Car Lunokhod 1 travels 6.5 miles (10.5 km) on the surface of the Moon.

December 1972 *Apollo 17* (USA) the last time man stands on the Moon.

▼ The Saturn V rocket was first fired successfully in 1967, and was specifically designed for the USA's manned space program. Here the rocket is lifting off from the pad at the Kennedy Space Center, Florida. The rocket's overall height is 363 ft (111 m) and at liftoff it weighs 2,850 ton (2,850 tonnes), powerful enough to launch a 150-ton (150-tonne) **payload** into Earth orbit and send a 50-ton (50-tonne) spacecraft to the Moon.

Man on the Moon

▶ Charles Conrad and Alan Bean landed their Lunar Module on Oceanus Procellarum on November 19, 1969. Their Lunar Module was only 200 yds (180 m) away from the Surveyor 3 spacecraft that had landed some 31 months previously. They did not want to land any closer in case their descent engine blew dust over the spacecraft. During the second of their four-hour sessions of extravehicular activity they walked over to Surveyor 3. Here Conrad is photographing Bean as he removes the TV camera. This camera and the soil scoop were brought back to Earth so that the effect of 31 months' exposure to the lunar environment could be assessed. The camera mirror was found to be pitted with microcraters and the spacecraft was covered with a fine, brown coating, masking its brilliant white paint. It had been sprayed by dust kicked up during its own descent.

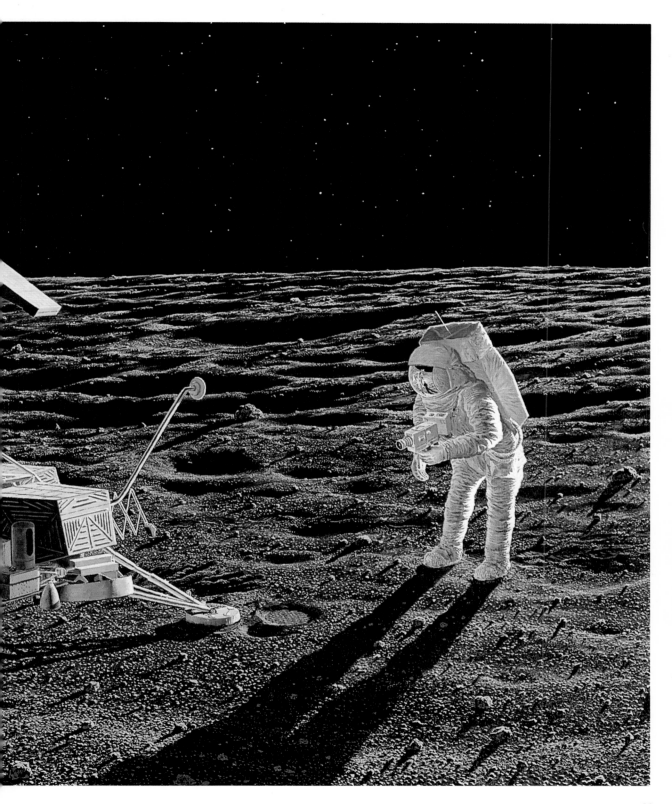

Apollo 17, the last visit

Neil Armstrong and "Buzz" Aldrin spent only 2 hours 12 minutes outside their Apollo 11 Lunar Module. They also walked a mere 550 yds (500 m). The Apollo 17 astronauts Eugene Cernan and Harrison Schmitt landed near the crater Littrow in Mare Serenitatis and spent 22 hours in total outside the Module. In their lunar roving vehicle they traveled a distance of 19 miles (30 km).

Rovers, transported to the Moon in a collapsed state, were also used on Apollo 15, 16 and 17. As the rover was deployed from its storage bay on the Lunar Module it slowly unfolded, its joints and wheels clicking into place. The lunar rover was powered by two 36 V silver zinc batteries, which drove four independent one-quarter horse power motors on each wheel.

In the Moon's low gravitational field the rover weighed only 77 lb (35 kg). On board was a TV camera and a very sensitive antenna to enable Mission Control on Earth to monitor the astronauts' movements. Two suited men could be carried easily, together with a whole range of tools and bagged and documented rock and soil samples. The Apollo 17 astronauts brought back 245 lb (110.5 kg) of lunar material.

▼ Harrison Schmitt, a geologist, and the only scientist to go to the Moon, is on the left of the huge fragmented boulder. The boulder has rolled almost 1 mile (1.6 km) down the North Massive slope to its present resting place. The scratch marks on the boulder have been produced by Schmitt's sampling scoop. The lunar rover is parked on the right.

The mountains seen in the far distance, across the Taurus Littrow valley, underline the fact that most Moon mountains have gentle slopes.

Eugene Cernan took a series of photographs on the Moon and this mosaic has been obtained by piecing them together.

Moon rocks

The rock samples have been analyzed by several laboratories on Earth and have revealed many unexpected details about the Moon. Analysis of the lunar rocks has enabled scientists to date specific incidents in the Moon's history.

The rocks on the Moon have been found to be chemically different from Earth rock and meteorites. The major difference is that Moon rocks contain no water, while Earth's normally contain between 1 and 2 percent. The Moon rock was formed from cooling lava, but, unlike on Earth, there was very little free oxygen. The samples brought back to Earth are stored in a dry nitrogen atmosphere to keep them from rusting.

There are two types of lava, the dark titanium-rich basalt lava of the mare and the light-colored highland rock which was produced in the early days of the Moon's evolution as a "skin" on top of a surface of molten rock. In general, the Moon's soil seems to have lost the relatively volatile (gaseous) elements present on Earth. It has also lost much of its iron and the elements that are often found with iron, like nickel and cobalt, but aluminum, calcium, titanium, barium, strontium and uranium are more prevalent than on Earth.

Billions of years of meteoritic bombardment have pulverized the Moon's surface. About 1 percent of this meteoritic debris can still be found in the soil. The top 0.5 in (1 cm) contains many very small glass beads produced as the soil melted during crater formation.

Footsteps in the dust

Man's last footsteps on the Moon were planted in December 1972. The Apollo missions had been a huge success. We had been peering at the Moon from our atmosphere-shrouded planet for many centuries. Though the telescope revealed a host of details, even from 238,900 miles (384,400 km) away, only our visits to the Moon have proved that it is a small, rocky, rubble-strewn world. There is no water or atmosphere and a large amount of dust. A manned base could be established there, even if we would have to live underground to compensate for the variation in temperature.

We are now ready for more distant trips. The Moon missions have provided a stepping stone to the next stage of our exploration of the solar system. They proved that it is possible to travel through space and land on other planets, to live on those planets for a time, and then return to Earth. Exploration of the solar system has come closer to being a reality.

Although we have learned an enormous amount about the Moon and the history of the processes which fashioned its surface, its exact origin still remains veiled in mystery.

Glossary

accretion: the growth of an object by the coming together of many smaller bodies.
basalt: a dark gray or black fine-grained rock formed when magma solidifies.
concentric: circles having the same center.
corner cubes: if the inside of a corner of a cube is given a mirror coating, light entering it will be reflected back the way it came.
density: the mass of a substance divided by its volume (i.e. iron is twice as dense as rock and rock is 3.3 times as dense as water, so water has a lower density than rock).
differentiation: a geological term: when a planet heats up and becomes molten the dense iron and metal falls to the center and the lighter rocks float to the surface. The planet has been differentiated.
dynamo: a region where moving electrically-charged particles produce a current which itself produces a magnetic field.
ejector: material blown out of a crater during its formation.
electrostatic: when substances become electrically charged, the positively-charged objects repel other positively-charged objects and attract negatively-charged ones.
escape velocity: the minimum speed an object must have if it is to escape from the gravitational attraction of a planet, satellite, or other body.
ghost crater: a crater that has been flooded with lava so that only the highest ring of mountains can be seen peeping above the lava.
genesis rock: the origional rock that accreted to form the Moon.
gravitational field: the space surrounding a body within which its gravity affects other bodies.
hypervelocity: a velocity that is considerably greater than the velocity of sound in the substance.
inert gas: a gas made of atoms like argon, helium, xenon, etc., atoms that do not combine with others to form molecules.
liquefy: to convert a solid substance into a liquid.
magma: liquid rock that flows out from under the surface of a planet or satellite.
magnetic field: a region where a magnetic force is acting.
magnetometer: an instrument for measuring the magnetic field.

mascon: a region of high density on the lunar surface which produces an increase in the gravitational field in the locality.
mass: the quantity of material in an object. (The property of an object determining the acceleration that would be produced by the application of a force.)
nebula: a tenuous cloud of material (usually gas and dust).
new Moon: only the edge of the Moon is lit by the Sun and appears as a thin crescent seen from Earth.
parking orbit: an orbit close to Earth where spacecraft "park" until conditions are suitable for them to be fired across space.
payload: the part of the total mass of a launch vehicle additional to that needed for its operation, for example, astronauts, passengers, cargo, mail.
radial: moving away from a point, like the spokes of a wheel.
radiate: to send out energy as light, heat or radio waves.
radioactive decay: an atomic nucleus decays into another species by emitting a much smaller atomic particle. Energy is also produced by this process.
rill: a V-shaped valley that winds across the surface of the Moon.
seismic: relating to the vibrations caused by earthquakes.
stage: most rockets have a series of stages which are discarded as the rocket moves away from Earth. The Saturn V rocket had three.
thermoelectric generator: a device for converting heat energy into electrical energy.
thrust: the force that can be generated by a rocket.
tides: the distortion of a planet's, or satellite's, surface by the gravitational pull of a nearby object.
velocity: the distance moved by an object in a certain time, for example, 10 miles a second (16km a second).
viscosity: the more liquid (runny) a substance, the higher is its viscosity.
volatilize: convert from a liquid into a gas.
water of crystallization: rock often contains water molecules in the form of solid crystals. When the rock is heated, usually to above 830°F (500°C) this water is released.

◀ The Sun and the nine planets compared, with their sizes depicted on the same scale. In comparison to the planets, the Sun is so large that only a small part of it can be shown.

Index

A
accretion, 6
Aldrin, Buzz, 35, 40
Alps, 23
ALSEP, 36
Apennines, 23
Apollo (basin), 28
Apollo mission, 24, 33, 34, 35
   36, 38, 40, 41
Archimedes (crater), 24
Aristarchus (crater), 24
Armstrong, Neil, 35, 40
asteroids, 12, 13, 17, 18, 22,
   23, 24
asthenosphere, 31
atmosphere, 18

B
basalt, 20, 41
Bean, Alan, 38

C
central peaks, 14, 18, 22, 32
Cernan, Eugene, 40
collision velocity, 16
Command Module, 34, 35
Conrad, Charles, 38
continental drift, 31
Copernicus (crater), 24, 32,
   33, 34, 35
core, 6, 31
crater, 14, 17, 18, 20, 22
   multiringed, 27, 28, 29
crater counting, 24
crater rings, 17, 27
cratering rate, 27
crust, 11, 20, 22, 28

D
dating, 24
depth to diameter ratio, 22
Descartes Highlands, 35
differentiation, 28
domes, 22

E
ejector, 17, 18
ejector angles, 18
Eratostenes (crater), 22
escape velocity, 8

F
far side of the Moon, 20, 28
Flamsteed (ghost crater), 24
Frau Mauro, 35

G
Gemini mission, 34
genesis rock, 12
ghost craters, 24
Giordano Bruno (crater), 27
gravitational field, 8, 17
Grimaldi (crater), 27

H
Hertzprung (basin), 28
highlands on the Moon, 20,
   28
hypervelocity impacts, 18

I
Iapetus (large moon of
   Saturn), 17
Imbrium Basin, 12, 13, 14, 16,
   18, 20
impactquakes, 31

K
Kennedy, John F, 32
Kepler (crater), 24

L
lava, 11, 20, 22, 24, 41
light flash, 18
lithosphere, 20, 31
Littrow (crater), 40
Luna, 20, 36
Lunar Module, 34, 35, 38
Lunar Orbiter, 33, 36

M
maria, 11, 20
mascons, 20, 27
Mercury, 17
Mercury mission, 34
magnetic field, 31, 36
man on Moon, 32
Messier twins (craters), 23
modification, 18
Moon
   atmosphere, 6, 7, 8
   birth, 7
   color, 30
   density, 6
   distance from Earth, 6
   inclination, 6
   map, 31
   mass, 6
   rock, 24, 41
   spin rate, 6, 7
   surface gravity, 6
   temperature, 6, 28
moonquakes, 31, 36

N
North Massive, 40

O
Orientale Basin, 18, 27, 28
origin of Moon, 12, 13

P
phase effect, 30
Plato (crater), 24
Procellarum, Oceanus (Ocean
   of Storms), 24, 27, 35, 38

R
radioactive decay, 24
Ranger, 32, 36
Riccioli, Giovanni, 23
rift valleys, 22
rills, 22
roving vehicle, 35, 40

S
saturation bombardment, 24
Saturn IV B rocket, 35
Saturn V rocket, 34, 37
Schiller (crater), 23
Schmitt, Harrison, 40
secondary craters, 14, 18
Serenitatis, Mare (Sea of Serenity), 20, 35, 40
Service Module, 34, 35

shock waves, 8, 18
soil, 22, 28, 30, 33, 35, 41
solar wind, 36
Sputnik, 32
Surveyor, 33, 36, 38

T
Taurus Littrow (valley), 40
terraces, 18
tidal bulge, 13

tides, 12
Tranquillitatis, Mare (Sea of Tranquility), 35

V
volcanic activity, 31

W
water of crystallization, 8
wrinkle ridges, 22

Acknowledgments

ILLUSTRATIONS
COVER, 7, 11, 13, 15, 17, 19, 20, 21, 23, 25, 28, 29, 31, 32, 33, 35 bottom, 37, 38, 39, 40, 41: Don Davis.
6, 8, 12, 16, 22, 34: Sebastian Quigley/Linden Artists.

PHOTOGRAPHIC CREDITS
9: NASA/BLA. 30 bottom: John Sandford/Orange, California, USA. 35, top: R.M. McNaught/Anglo Australian Observatory, Sliding Spring, New South Wales, Australia. 43: BLA.